CW00531562

ICONIC AND UNIQUE –
MEET THE NEW MERSEY GATEWAY BRIDGE

FOREWORD
A FEW WORDS FROM ROB POLHILL, LEADER OF HALTON BOROUGH COUNCIL

/// *When Her Majesty The Queen officially opened the Mersey Gateway on 14 June 2018, it symbolised the end of an epic 20-year journey and the start of a bright new era for Halton.*

At that truly memorable event, which was attended by many of those who had worked so hard over the years to deliver the project, I was reminded of the many challenges we faced in building this bridge: of the years when our pleas for a new crossing were dismissed; when, tantalisingly close to securing funding, a general election, ministerial change then a spending review frustrated our progress; and how the formidable River Mersey defied our engineers to push boundaries to find new construction solutions.

There were days when it seemed that our dream would never become a reality. But we persevered and today, the magnificent, elegant Mersey Gateway Bridge, stands proudly before us bearing testament to the fact that partnership, hard work, resilience and determination will always succeed.

This book tells our story. One of challenge and triumph which, through some wonderful photographs, provides an insight into this once in a generation project. It shines a light on some of the unique and fascinating aspects of the scheme, the people involved and the impact of the Mersey Gateway Project across the community.

From the outset we said that the Mersey Gateway would be more than just a bridge, that it would be a bridge to prosperity, and we are already seeing this become a reality. The bridge is cutting journey times and providing a more reliable route. Jobs and apprenticeships have been created for our young people directly by the project and through increased levels of investment in the borough. New opportunities for development and regeneration are creating possibilities for further growth that will see Halton continue to thrive in the years to come. The future is indeed exciting.

It has been my absolute privilege to have been involved with the Mersey Gateway Project. I would like to give my heartfelt thanks to my predecessor Cllr Tony McDermott MBE and to Derek Twigg MP for their invaluable contribution over the years, as well as to each and every other person who played a part in getting us here.

I'm delighted to be able to share this book with you, I hope you enjoy reading it.

Cllr Rob Polhill
Leader of Halton Borough Council

CONTENTS

LOCAL WORKERS AND SPECIALISTS

FROM AROUND THE WORLD
ALL PLAYED THEIR PART

THE CONSTRUCTION OF THE
MERSEY GATEWAY BRIDGE
AND CONNECTING IT TO
THE ROAD NETWORK

WAS AN INCREDIBLE FEAT
OF ENGINEERING

PART 1:
BUILDING THE MERSEY GATEWAY

GETTING TO THE START LINE

/// *It took a year to build a temporary access bridge – the trestle bridge – just so we could start work in the middle of the river.*

THE 7 MAY 2014 MARKED DAY ONE OF CONSTRUCTION, BUT THE NEW BRIDGE WAS A LONG WAY OFF.

A major part of the project was the extensive programme of temporary works needed in order for the construction of the bridge and its approach structures to begin.

Just getting vehicles and equipment to the river was a task in itself. The river works were undertaken initially by use of barges however, due to the ever changing tides in the estuary, a new approach was needed.

The team spent a year building a temporary access bridge – the trestle bridge – across the River Mersey between Widnes and Runcorn.

A spectacular feat of engineering, the temporary trestle bridge was the first bridge built across the River Mersey in Halton since the Silver Jubilee Bridge opened 53 years earlier.

The trestle bridge was 1,000m long, but just 9m wide, and acted as an access platform for construction teams to work from when building the main bridge.

It had local roots – 200,000 tonnes of locally sourced stone was used during its construction and the hundreds of concrete slabs which formed the road surface were made in Northwich in Cheshire, while the metal frame was manufactured in Warrington.

A COMING TOGETHER:

ON THE 31 MAY 2015, THE PROJECT CELEBRATED REACHING ITS FIRST MAJOR MILESTONE WHEN THE NEW TRESTLE BRIDGE WAS COMPLETED

Abbie Strain from Pewithall Primary School in Runcorn, captured the mood:

/ / / *It's amazing to be out here in the middle of the River Mersey – it's so big. The view you get is really impressive and I'm really pleased to help open this temporary bridge. I can't wait to see how the big new bridge looks while they are building it.*

Students from Pewithall Primary School in Runcorn, and Widnes Academy, were the stars of the show when the trestle bridge was completed in May 2015.

The children met in the middle of the Mersey – with TV cameras in tow – took some amazing selfies, swapped school 'bears' and made special hand imprints from clay to represent the building of the bridge.

DIGGING DEEP IN THE RIVER MERSEY

The cofferdams were like three giant, dry, watertight steel circles in the middle of the river. It's an amazing place to work.

Work began in the middle of the estuary to create the three temporary cofferdams needed to allow construction of the three bridge pylons.

The three cofferdams created huge dry working areas – in the shape of giant circles – so the team could lay the concrete base for the pylon foundations.

This was done by driving steel sheet piles into the riverbed to form two enclosed circles at each location – an outer circle of 40m in diameter and an inner circle of 20m in diameter. Almost 300 steel piles were used for each cofferdam, with each one driven 12m into the riverbed.

Once they were built, the water was pumped out and the outer areas were filled with local natural materials, such as stone or sand, to create a dry environment.

A concrete floor was laid inside each cofferdam, and a cage made of steel reinforcing bars was assembled.

Upright steel reinforcement bars were then fixed into the centre of the cages to form the beginnings of the three pylon shafts.

Thousands of cubic metres of concrete was carefully poured into each cofferdam to form the foundations of the pylon shafts – these are the tall structures that you can see today rising out of the riverbed to support the bridge deck.

If that sounds complicated, then remember the River Mersey was rushing past just outside.

The Upper Mersey Estuary, which takes in the Silver Jubilee Bridge and the Mersey Gateway Bridge, has two distinct river channels and is hugely influenced by the tides that come in and go out every day. Freshwater from the River Mersey mixes with the saltwater coming in with the tide from the Irish Sea, covering and exposing the riverbed on a regular basis.

It was important for the project to ensure that this natural environment surrounding the bridge was preserved and improved where possible.

The need to avoid artificially affecting the naturally random behaviour of the river channels heavily dictated the locations of the cofferdams and the design of the actual bridge.

TRINITY AND WEBSTER – OUR BRIDGE BUILDING MACHINES

The bridge is actually made up of three parts – the main bridge structure across the river and the two elevated approach viaducts connecting it to the main road networks in Runcorn and Widnes.

Today, 20 supporting piers hold up the viaducts that take traffic across the saltmarsh to and from the Mersey Gateway Bridge. It took around two years to put them in place and to build the road deck that connects them.

In order to do that, the team had the help of Trinity and Webster – two huge and very complex pieces of engineering machinery.

Trinity and Webster were bespoke Moveable Scaffold Systems (MSS), specially designed to construct the curved north and south approach viaducts respectively leading to the Mersey Gateway Bridge. These viaducts run across the Mersey saltmarshes and span the St Helens Canal in Widnes and the Manchester Ship Canal in Runcorn.

Trinity was named after the three bridge pylons and organisations which make up the Merseylink Consortium. Webster was given its name by Halton schoolchildren and pays homage to local engineer John James Webster, who designed the Widnes Transporter Bridge.

Both machines were 157m long – the length of one and a half football pitches – 8m high and up to 22m wide, each weighing approximately 1,700 tonnes – the same as 140 double decker buses. These were the two largest MSS type machines working in the world at that time and the first time that an MSS has been used for such a major viaduct.

Trinity and Webster acted as giant concrete moulds which were locked onto the bridge piers before pouring concrete into the mould to create a deck span. Once the first span of the viaduct was complete, the machine moved along via hydraulic jacks to create the next deck span and so on until all eleven spans of the north approach and eight spans of the south approach were complete.

These remarkable machines meant there was just a 21-day cycle time for construction of each 70m long span. The machines allowed 1,250m^3 of viaduct bridge deck to be constructed with each span being a single pour.

Between them, Trinity and Webster used 27,600m^3 of concrete, enough to fill more than eleven Olympic size swimming pools.

Once Trinity and Webster finished work, special 'wing travellers' were then used to build the outer road lanes on either side of the approach viaducts.

Each wing traveller weighed 280 tonnes, was 48m wide and 20m tall. They worked in a similar way to the MSSs, acting as movable concrete moulds which completed the full 43.5m deck width, which carries six lanes of traffic.

Each machine was fixed onto two railway tracks that sat on top of the deck section that had already been cast by the MSS.

The Mersey Gateway Bridge may be built, and Trinity and Webster no longer needed, but both live on in spirit. Both MSSs were recycled and sent to Bratislava in Slovakia to help build a new bridge over the Danube.

TRINITY AND WEBSTER
WERE 157M LONG

8M HIGH AND
UP TO 22M WIDE,
EACH WEIGHING
APPROX 1,700 TONNES

WING TRAVELLERS
WEIGHED 280 TONNES,
WERE 48M WIDE AND
20M TALL

THE MERSEY GATEWAY BRIDGE

THE COMBINED LOAD BEARING ABILITY OF THE STAY CABLES IS 53,000 TONNES – THE EQUIVALENT OF 9,000 ELEPHANTS!

The design of the Mersey Gateway Bridge is unique. The three bridge pylons are all different sizes – 110m (north), 80m (central) and 125m (south).

Constructing them was no easy task. Specialist 'auto-climbing system formwork' was used on each pylon to make sure the work was done safely and effectively.

It worked by building a section of the pylon then repeatedly 'climbing' upwards to create the next section until the pylon was complete.

With the pylons in place and towering over the river, constructing the bridge deck was the next step.

Three pairs of 'form travellers' were used to build the bridge deck with each pair operating as a unit.

Designed and built specifically for the project and each weighing 270 tonnes, the six form travellers acted as moveable concrete moulds on the main bridge itself, operating in a similar way to Trinity and Webster.

They were lifted up to bridge deck height using hydraulic ramps, then the three pairs acted as cantilevered formwork to construct the one kilometre long bridge deck piece by piece. There are 154 deck segments, each one 6m long and 33m wide.

Each pair started at one of the pylons and moved outwards, so the bridge deck effectively 'grew' from each side of each pylon until it was all joined together.

As the deck grew, the 146 stay cables that connect the bridge deck to the pylons were put in place one by one.

Today, the stay cables give the bridge a stunning look, but their beauty hides their strength.

Each outer cable contains up to 91 steel strands, which laid together would stretch over 810 miles; around the same distance from Land's End to John O'Groats.

INTERVIEW:

LIFE AT THE TOP OF A TOWER CRANE

Merseylink's tower crane team played a key role in the construction of the three Mersey Gateway Bridge pylons, which now dominate the skyline in the Mersey Estuary.

The tower cranes were used for lifting materials and reinforcements for the bridge pylons as well as for constructing and dismantling the temporary scaffolding works that allowed construction workers to access the site.

The largest tower crane, which was located on the Runcorn side of the river, stood at 146 metres – over 479 feet – at its maximum height, the equivalent of 32 double decker buses stacked on top of each other.

We asked lead tower crane operator, Peter McDonough, a few questions to find out what it was like to work with your head literally in the clouds.

Peter, who has been driving tower cranes for 18 years, headed up a team of 12 highly skilled tower crane drivers, employed by MPS Crane Operators – a specialist lifting organisation that supplies professional crane drivers all over the world.

So Peter, what's it like working so high up off the ground?

Great. Quiet and peaceful with the most incredible views. The view from this job was the best I've ever had, you could see for miles.

The tower crane is nearly 500ft high – how do you get up there?

I climb up! There are around 520 steps and on a steady climb it can take around 25 minutes, but there are rest platforms where you can catch your breath.

Getting up to the cabin is definitely the hardest part of the job. It's not easy or pleasant on a cold, wet morning!

Do you ever feel scared up there?

You definitely need a head for heights to do the job but no, the height doesn't bother me at all. It's the safest place to be on a construction site. That's not to say it isn't dangerous but you're out of the way of everything.

What are the tower cranes being used for?

The tower cranes lift, transfer and place various materials and equipment such as reinforcement cages, formwork and the steel stay cables.

What does a shift involve?

Before starting work we have a pre-shift briefing at the base of the crane. Going up the ladder I do visual safety checks, checking the pins, bolts, structure and ladders on the crane. When I get into the cabin I complete a thorough safety checklist to make sure that everything is working correctly. I then radio down to confirm that everything is safe and operational and the day's work will begin.

What's the best thing about the job?

Being from Liverpool, I really wanted to get on the project as it's local to me and I knew it was going to be high profile. It's a fantastic job. The engineering expertise is just phenomenal and we have a great team of people that support one another.

KEEPING HALTON ON THE MOVE

/ / / *One of our biggest challenges was keeping traffic moving during three and a half years of construction right across Widnes and Runcorn. People were incredibly patient and the team owes them a huge thank you.*

The Mersey Gateway Project has always been about much more than building a bridge.

THERE WERE 12 NEW BRIDGES ACROSS THE PROJECT, AND EIGHT NEW OR MAJORLY IMPROVED ROAD JUNCTIONS.

It also included 9.2km of new or improved road network, connecting the new bridge to the M62 to the north and the M56 to the south.

As a result, the road network around Halton today is unrecognisable in parts from pre-Mersey Gateway construction – and is much better for it.

Major new junctions were created at Ditton (Widnes) and Bridgewater (Runcorn), with remodelling of other main routes including the M56 Junction 12 and the Central Expressway. Elsewhere, demolition of old roads and bridges took place to make way for new infrastructure.

Over three and a half years this work progressed, while those living and working in the busy towns of Runcorn and Widnes went about their daily lives.

Inevitably, with an infrastructure project of this scale, there was disruption, with local people and businesses bearing the greatest burden.

The scale of planning and communications work involved to ensure that the region was kept on the move was huge. Every week a real effort was made to keep people informed, with frequent, honest and transparent communication at the heart of the complex traffic management strategy.

TWO MINUTES TO MIDNIGHT – THE OPENING NIGHT

/// We had the most amazing fireworks launched from the deck of the Mersey Gateway Bridge. Then, we had two hours to clean up, close the Silver Jubilee Bridge at midnight and just a minute later open the new Mersey Gateway Bridge and keep everyone moving. It was a very busy night.

THE MERSEY GATEWAY BRIDGE OPENED ON TIME AND TO BUDGET AT MIDNIGHT ON SATURDAY 14 OCTOBER 2017.

The historic event was celebrated across the community with a spectacular light and fireworks display launched from the new bridge at 9pm. Tens of thousands of people gathered on both sides of the river to be part of local history and the accumulation of a 20-year journey of challenge and triumph to watch the iconic Mersey Gateway Bridge open to traffic.

Despite the midnight opening time, thousands of motorists, bikers and camera crews all did their best to be the first across the bridge and the opening went without a hitch.

THE MERSEY GATEWAY BRIDGE

OPENED ON TIME AND TO BUDGET

AT MIDNIGHT ON SATURDAY
14 OCTOBER 2017

the **mersey** gateway
A BRIDGE TO PROSPERITY

OFFICIALLY OPENED BY

Her Majesty The Queen

ON

14TH JUNE 2018

A CEREMONY FIT FOR A QUEEN (AND A DUCHESS)

*/ / / Her Majesty and the Duchess both seemed to really enjoy the performances –
and the children were so excited – it was a wonderful day.*

ON 14 JUNE 2018, HER MAJESTY THE QUEEN, ACCOMPANIED BY HER ROYAL HIGHNESS THE DUCHESS OF SUSSEX, VISITED HALTON TO CELEBRATE THE OFFICIAL OPENING OF THE MERSEY GATEWAY BRIDGE.

The outing was Her Majesty and Her Royal Highness' first ever engagement together and they both appeared delighted by the magnificent new bridge.

Pictures copyright Halton Borough Council

Your Majesty, Your Royal Highness, distinguished guests; we are delighted to welcome you to Halton today, to officially open our magnificent Mersey Gateway Bridge.

Cllr Rob Polhill, Leader of Halton Borough Council

The visit began with the arrival of the Royal Train at Runcorn Station where Her Majesty The Queen and Her Royal Highness were greeted by local school children and members of the public who had lined the nearby streets.

The pair then travelled through Halton, crossing the Mersey Gateway Bridge to the grounds of the Catalyst Science and Discovery Museum in Widnes, where the official opening event took place in full view of the iconic bridge and over 600 local primary school children.

Her Majesty The Queen and Her Royal Highness were treated to a specially commissioned performance entitled "Bringing Communities Together: Bridges of Halton" which was performed by 80 local school children, and directed by the locally based Andrew Curphey Theatre Company.

The moving piece told the story of the many crossings that have existed over the Mersey connecting Widnes and Runcorn through the ages.

THE OFFICIAL OPENING EVENT
TOOK PLACE IN FULL VIEW OF THE ICONIC BRIDGE AND OVER 600 LOCAL PRIMARY SCHOOL CHILDREN

The event was attended by many of the people who helped make this bridge a reality – our local politicians, our partners, staff, construction consortium, operating teams and volunteers.

Four lucky children also played a special role in the event, each presenting Her Majesty and Her Royal Highness with a posy of flowers. They were chosen because their birthdays fell on 14 June (the official opening date) and 14 October (the date the bridge opened to traffic).

Pictures copyright Halton Borough Council

GET IN LANE

(M62)
Widnes (A557)
St Helens (A570)
Warrington (A562)
↓

½ m

GET IN LANE

(M62)
Widnes (A557)
St Helens (A570)
Warrington (A562)
↓

(M57) Liverpool & ✈ A533 (A562)
↓   ↓

60

PAY TOLLS
ONLINE
AT
MERSEY LOW

HIGHWAY MAINTENANCE

QUICKER, EASIER AND MORE RELIABLE JOURNEYS

In the first year the bridge has welcomed over 23 million journeys at an average of 50 miles per hour.

The Mersey Gateway Bridge uses state of the art technology to allow people to cross the river without having to stop to pay for their crossing.

This free-flow tolling system means there are no toll booths – making for quicker, easier and more reliable journeys.

Free-flow tolling is widely used across the globe and across several stretches of road in the UK and Ireland. Emovis, the company who run the tolling operation on the new Mersey Gateway Bridge (and later on the Silver Jubilee Bridge) under the name Merseyflow are global leaders in the field.

The tolling gantry spans the Widnes side of the bridge, tracking every vehicle that crosses.

As traffic uses this free-flow system, dedicated cameras and sensors on the gantry read vehicle number plates and special Merseyflow stickers installed on the windscreens of registered vehicles, to identify them.

As drivers travel under the gantry a camera detects the vehicle's front and rear number plates. A separate sensor can also detect whether a vehicle has a Merseyflow sticker installed and, if so, scans this and records the journey.

Information collected by the cameras and sensors is encrypted and sent to Merseyflow's offices, where it is used to charge customers for their crossings and to analyse traffic flow.

Even before the bridge opened, the cameras were being put to use as a series of lorries, cars and other vehicles were driven across the bridge to test the system before it opened.

Around two million vehicles a month are using the bridge – all passing under the gantry and being monitored by the cameras and the sensors – making it one of the most sophisticated tolling systems in the world.

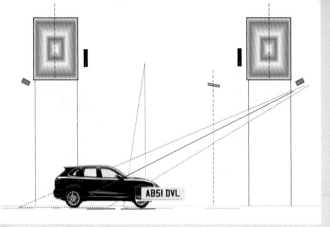

23.25 MILLION CROSSINGS
IN THE FIRST YEAR OF OPERATION

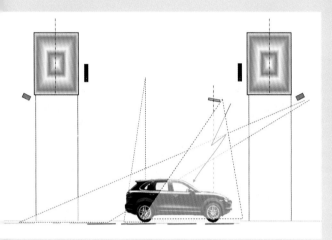

72,000 CROSSINGS
PER WEEKDAY ON AVERAGE

2,070,000 CROSSINGS IN A SINGLE MONTH (JULY 2018) – **THE HIGHEST NUMBER OF JOURNEYS RECORDED** SO FAR

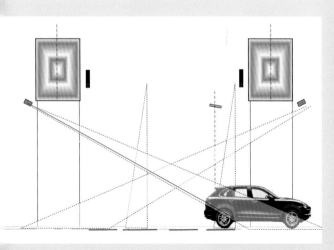

291,000 PEOPLE HAVE REGISTERED FOR DISCOUNTS WITH MERSEYFLOW

63,000 PAYMENT TRANSACTIONS BEING **COMPLETED WEEKLY**

REDUCTIONS IN JOURNEY TIMES OF
UP TO 20 MINUTES

PART 2:
COMMUNITY AND PEOPLE

This was an incredible opportunity for us to really get young people excited by the amazing engineering, design and construction processes.

INSPIRING THE
NEXT GENERATION

This was an incredible opportunity for us to really get young people excited by the amazing engineering, design and construction processes.

Across many generations, the River Mersey and the transport history of Halton has shaped how families in Runcorn and Widnes live, work, travel and socialise. It was important to make sure that young people and those in education were involved in this moment of history, just as their families were before them.

Through all stages of the project, the team aimed to ensure that local school children were not only aware of the changing landscape of Halton, but had the opportunity to be actively involved; learning and making memories with the Mersey Gateway.

LESSONS IN BRIDGE BUILDING
BECAME PART OF HALTON'S PRIMARY SCHOOLS' CURRICULUM

School children were with us when the temporary trestle bridge came together in the centre of the River Mersey in May 2015, when we opened both of the project's visitor centres, when we opened the brand new Lodge Lane footbridge in Runcorn and of course when the opening of the Mersey Gateway Bridge was celebrated.

Inside school, lessons in bridge building became a part of Halton's primary schools' curriculum and the dedicated Mersey Gateway volunteers delivered presentations to 2,200 children across 71 primary and secondary schools.

The lessons about bridge building (with use of Lego), environmental impacts of construction, construction site safety and civil engineering careers were a huge hit!

THE CREATIVE CONNECTIONS SCHEME WAS A BIG SUCCESS

Arts initiatives made up another large part of the project's work with schools. The Mersey Gateway Project's Creative Connections scheme partnered teachers from 36 Halton primary schools with local, professional artists and was a big success.

The programme provided an avenue for pupils to develop creative work, with the support of a local artist, on the themes of environment, local history, and bridge design and construction.

THE TIME BANK COMMUNITY

/// *Time Bank is all about the whole team giving something back to the local community. It's been incredibly rewarding and actually quite emotional at times.*

From day one the project team has been dedicated to providing a helping hand to local communities right across Halton.

The Time Bank scheme set up by Merseylink was at the heart of that – the Merseylink team and its sub-contractors provided free practical help and support for local organisations, community projects and educational establishments.

In total, over 20,000 hours of support was delivered, with £121,000 worth of community benefits delivered locally across 18 projects in Runcorn and Widnes.

The scheme worked by matching successful applicants with relevant partners within Merseylink, who volunteered time and resource to provide specialist skills and assistance. Expertise on offer included strategic development, financial and commercial advice, as well as help from construction, engineering and environmental specialists.

Schools, sports clubs and community centres all benefited.

Jill Berry, Manager of Halton Lodge and Grange Pre-School, was thrilled with the final result.

/// *Our Ofsted inspector even made some positive comments about our school environment. We successfully maintained a grade of 'Good' and I believe that the works carried out by Merseylink really contributed to this rating by enhancing the provision that we offer.*

Chris Wheatley, Captain of Runcorn Rowing Club, said:

/// *This magnificent gift of time and expertise enabled us to install our new clubhouse and teaching facility. It has made an enormous contribution to our club and I would like to thank the Mersey Gateway team for their support.*

Jean Wright, great grandmother of Oliver Wright, age two, said:

/// *I've lived in this community for 53 years and this new play area is going to make an amazing difference. The children now have somewhere to play safely, helping them to get fresh air and exercise.*

JOBS AND SKILLS FOR THE FUTURE

/ / / A vision for the Mersey Gateway was to ensure that the project was so much more than just a bridge.

Halton Borough Council believed the positive impacts of the new bridge and road network should reach as many people as possible.

Right from the beginning the project was seen as a catalyst for new jobs, learning and employment opportunities.

The Council wanted to encourage and include a strong local workforce and incorporated a robust employment and skills element into the selection process when choosing a consortium to deliver the project.

When the Merseylink Consortium was announced as the preferred bidder, its proposals, such as a dedication to creating jobs, apprenticeships and work placement opportunities became part of the contract.

This commitment was backed up through the Merseylink supply chain, with principal and sub-contractors all playing their part.

The backbone of this project was its people and, across the three and a half years of construction, almost 5,000,000 man hours were spent working on site, and more than 25,000 people from at least 30 countries across the globe have worked on the project.

At the busiest period around 1,300 people were working on site right across Runcorn and Widnes.

The project team encouraged and created opportunities for apprentices and work placements, seeing 119 days of work placement visits by unemployed Halton residents, 32 apprenticeships in areas from accountancy to quantity surveying and a total of 73 weeks' worth of work experience offered during construction.

A series of pre-employment courses were developed by Merseylink, including a special training course involving Merseylink, Riverside College in Widnes and Halton Employment Partnership, helping local residents gain the knowledge and practical skills required for a career in construction.

The specially created six-week courses included four weeks spent with Halton Employment Partnership where trainees were given help and advice to prepare them for the workplace. This included confidence building, teamwork, CV writing, mock interviews, and presentation tips.

Trainees then attended Riverside College for a week, where they had an introduction to working on large construction sites, including health and safety and practical construction training, with the final week based on site with Merseylink. This involved a health and safety induction, a site tour and on-the-job training with the bridge-builder's specialist teams.

At the end of the course, trainees were offered the opportunity to apply for a construction apprenticeship with Merseylink.

During the construction phase, 17 Halton Employment Partnership pre-employment programmes were delivered, supporting over 100 residents in learning new skills, many of whom secured an apprenticeship or employment with Merseylink or its supply chain.

Over 650 people from the Halton area were successful in securing employment on the project. These included roles in general construction, traffic management, administration, accounts, IT, quantity surveying, and steel fixing.

AT THE BUSIEST PERIOD AROUND 1,300 PEOPLE
WERE WORKING ON SITE RIGHT ACROSS RUNCORN AND WIDNES

WOMEN IN ENGINEERING

The construction industry has often been seen as a 'man's world', Mersey Gateway is proof that this is definitely not the case.

Women from around the world have been involved in helping construct the Mersey Gateway Bridge and re-designing and rebuilding the approach roads to connect it to the regional road network.

They have taken on key roles, working across the project in civil engineering, health and safety, environmental, and administrative roles, with even more women joining the team through work experience opportunities and apprenticeships.

The team celebrated 'International Women in Engineering Day' each year to focus attention on the amazing careers available in engineering and technical roles, and celebrate the achievements of outstanding women engineers.

Jane Burgess, a Trainee Engineer who worked with Merseylink while studying for a degree in Civil Engineering at Leeds Beckett University said,

For me, working on the Mersey Gateway Project has been an incredible opportunity. My message to anyone thinking of developing a career in civil engineering would be go for it – lots of women are put off by thinking it is a male dominated industry, but there are lots of women working in a variety of roles on the project and it is an incredibly rewarding job. The satisfaction you get from being able to stand back and see something you have helped to build is amazing.

SHARING OUR STORY

To have over 30,000 people – from right around the world – come to our visitor centres is incredible. Our volunteers had a lot of fun telling them all about the bridge.

To help tell the story of Halton's new bridge and to allow the local community to learn more about the construction process, two visitor centres were developed, one on each side of the river.

The Widnes visitor centre at the Catalyst Science and Discovery Museum was opened on 10 February 2015 with the help of pupils from St Michael's Primary School, Widnes.

A second information centre was opened in Runcorn Shopping City in February 2016 with the help of pupils from St Mary's CE Primary School, Runcorn. This centre closed to make way for a new retailer to the shopping centre in August 2017 after welcoming 10,000 visitors.

Over 30,000 people from all across the world visited the two Mersey Gateway visitor centres during construction and the number continues to rise.

The two centres were staffed by Mersey Gateway volunteers, who acted as ambassadors for the project and key points of contact for the public.

Visitors were able to use a number of interactive display screens to access a wide range of project information, including videos, photographs, a live site webcam and a special children's zone. E-learning packages and training materials were also available, providing detailed information about construction techniques.

Volunteers also hosted trips to the Catalyst Science and Discovery Museum rooftop observatory, where visitors could get a stunning bird's eye view of the construction work which was taking place in the estuary.

The visitor centre in Widnes remains open and now holds both construction and environmental information with additional environmental tours around the Spike Island open space on the banks of the river.

Named as the 25,000th visitor, Arthur, aged 84, is pictured with his son Stephen on one of the bridge's pylons during construction.

OUR AMAZING VOLUNTEERS

/ / / *Our volunteers have been just amazing. They are special people who selflessly give up their time to share their knowledge of this fascinating engineering project with visitors and the people of Halton. In doing so, they are making an enormously valuable contribution, not just to the project, but to the local community as a whole. Thank you to each and every one of you.*

Hugh O'Connor, General Manager of Merseylink

Volunteers have been the beating heart of the Mersey Gateway Project. 92 local volunteers have and continue to make a massive contribution to sharing the Mersey Gateway story with young people and the local community.

A dedicated volunteer scheme launched in May 2014 with the aim of recruiting local people to be ambassadors for the new bridge project. It consisted of a four-week long training course where volunteers learnt presentation and customer service skills and discovered fascinating facts about the ecology, environment and local history and heritage of the project.

The volunteers then worked within the project's visitor centres to help explain how the bridge was being built and helped to work within the community to deliver talks, presentations and temporary exhibitions in local community venues, schools and colleges.

Throughout the construction period, all volunteers were recognised as part of annual award ceremonies where certificates, badges and mementos were presented to recognise their work and achievements – some were even recognised for over 2,500 hours spent in their volunteer role.

Volunteers continue to work with the project now the bridge is open to share the environmental story of the new bridge and to work within the local community and schools to raise environmental awareness.

INTERVIEW:

MEET EVELYN...

Evelyn Edwards is just one of the volunteers who made such a difference to the project. She explains what it meant to her.

Why did you decide to volunteer with the Mersey Gateway Project?

Several reasons. I wanted to be part of something that would go down in history, to meet like-minded people and have interesting conversations with them, and to make new friends and enjoy the social side of volunteering.

What does your role involve?

I volunteer at the visitor centre in the Catalyst Museum in Widnes. My role involves greeting visitors and explaining different aspects of the project. I talk them through various screens and presentations and take them up to the viewing gallery where they can see the construction in the river.

What do you like best about volunteering?

Meeting people from all walks of life, continually learning about what's happening at each stage of development, and watching visitors depart with a positive attitude and knowing that I played an active part in that.

Have you learnt any new skills?

I have learnt that becoming older does not mean that you have to start 'winding down' if you are not ready to.

What do you get out of volunteering on a personal level?

I get satisfaction out of educating visitors about the bridge. I have gained renewed confidence and I thoroughly value the friendships I have made with my fellow volunteers.

What is the most interesting or unusual question you have been asked?

If people will be allowed to climb to the top of the towers! The answer is no – unless you are a maintenance worker – we had to let people down gently!

Picture copyright Halton Borough Council

AROUND THE WORLD TO HALTON

/// *The interest we've had in the Mersey Gateway Project from around the world has been astonishing. People across the globe know about, and are learning from, this project.*

The project team knew that specialist workers on the project would come from around the world, but the distance that some people travelled to come and see the work was amazing.

Workforce

Visitors

Scandinavian students

/// *It was very interesting for us to see the machinery that was being used to build the bridge deck. Now we know how a bridge of this size can be built and the challenges that may be experienced during the construction process.*

Charles Hay, then the UK's Ambassador to South Korea, said;

/// *This project brought together experts from across the globe and it was fascinating to look at how the ideas and technology combine to deliver such a major piece of engineering.*

The South Korean Ambassador

The Shanghai Construction Group

The Bangladesh Army

Major General Abu Syeed Md Masud, from the Bangladesh Army, said:

/// *We were very pleased to have visited the site. We saw many things, which will be useful in our implementation of our bridges and roads in Bangladesh.*

THE MERSEY GATEWAY ENVIRONMENTAL TRUST

/ / / *What we're doing here in terms of bringing long-term environmental benefits to the area – alongside a major construction project – is really quite unique. It's special to be a part of it.*

A unique new charity – the Mersey Gateway Environmental Trust – was set up in 2010 to promote the conservation, protection and improvement of the environment across a 1600-hectare area of the Upper Mersey Estuary running all the way from the Silver Jubilee Bridge up river to Warrington.

The Trust worked in partnership with Merseylink during the construction stage to focus on the conservation of the environment and habitats surrounding the project route.

NOW THE NEW BRIDGE IS OPEN, THE WORK OF THE TRUST CONTINUES

THE TRUST IS EMBARKING ON A 25-YEAR JOURNEY TO CREATE

A NEW 28.5 HECTARE NATURE RESERVE

AROUND THE AREA OF THE NEW ICONIC BRIDGE

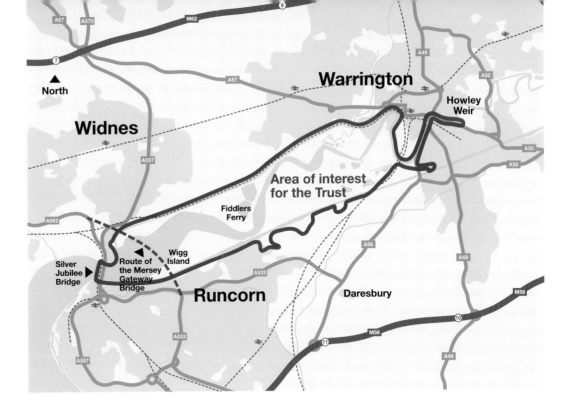

One of the Trust's key aims is to be the estuary's most active and influential wildlife champion and to ensure the project is a visionary example of environmental management for future major construction projects.

The Trust see the Upper Mersey Estuary as a 'living laboratory' and has supported research in many areas such as shrimp ecology, carbon sequestration, ecosystem services – all to bring about wildlife benefits to the estuary. The Trust also successfully introduced cattle grazing and began to manage reedbeds and saltmarsh habitats.

Already the Trust is active across the estuary. The Plastic Resolution project works with local community groups and school children and is designed to educate and take action about damaging plastic waste in our oceans and rivers. The equivalent of 420 bathtubs full of plastics has been removed from the Upper Mersey Estuary through the project so far!

THROUGH LOCAL FERRIES, THE LONDON TO LIVERPOOL RAILWAY, THE TRANSPORTER BRIDGE AND THE SILVER JUBILEE BRIDGE,

LINKS BETWEEN RUNCORN AND WIDNES STRETCH BACK 900 YEARS

THE CAMPAIGN TO GET A NEW BRIDGE HASN'T LASTED QUITE THAT LONG, BUT IT TOOK

20 YEARS OF HARD WORK

AND NOT TAKING NO FOR AN ANSWER FOR HALTON TO GET ITS NEW BRIDGE

THE MUCH-LOVED SILVER JUBILEE BRIDGE WAS SHOWING ITS AGE

AND BECOMING A REAL BARRIER TO ECONOMIC GROWTH IN THE NORTH WEST

PART 3:
WHERE DO YOU START?

This part of this book looks at the history of bridges in Halton and how Halton Borough Council achieved its aim of delivering the new Mersey Gateway Bridge.

HALTON'S BRIDGES: A BRIEF HISTORY

/// *Links between the two towns of Widnes and Runcorn stretch back 900 years to the early 12th century when land on both sides of the river formed part of the Halton Barony.*

Citizens from the Widnes area had to pay taxes at Halton Castle in Runcorn and a ferry service soon linked the two settlements.

Around the mid-19th century, the British chemical industry began in Widnes utilising Lancashire coal and Cheshire salt. Factories developed on both sides of the river and trade, labour and skills quickly became interlinked. As a result of this, in 1868 the London to Liverpool railway opened, followed in 1905 by a transporter bridge to carry road traffic and pedestrians.

These constructions reinforced the links between the two towns, which were increased even further when the Silver Jubilee Bridge opened in 1961. The steel arch bridge is listed for its architectural importance and was the third largest bridge of its type in the world when it opened.

A CONNECTED BOROUGH – SUPPORTING REGIONAL TRANSPORT

/ / / The Mersey Gateway Bridge means the area now has transportation links that are second to none.

Halton is located within the economic triangle formed by Liverpool, Manchester and Chester. Nowhere in Halton is more than 15 minutes' drive from a motorway junction.

Approximately a third of the UK's residential population and around a half of all British manufacturing business is located within a two hour drive.

Businesses have been attracted to the area by the numerous advantages that set Halton apart as a business location, including a good supply of suitable, value for money sites and premises,

appropriate skills and a pool of existing firms to act as suppliers and sub-contractors.

Halton is located between both Liverpool's John Lennon Airport and Manchester Airport. It also offers direct access to the West Coast Mainline for both rail freight and passengers. Proximity to the Liverpool and Birkenhead ports means that shipping to global destinations and ferries to Ireland and Isle of Man are easily accessible.

The Mersey Gateway Bridge means the area now has transportation links that are second to none, providing effective road connections to the Liverpool City Region from the south, thereby improving connectivity and reducing congestion.

Transport is an essential component of the Liverpool City Region's objectives for economic growth, skills, health, housing and regeneration as part of the continuing transition to a dynamic and growing low carbon economy.

The new crossing enhances the existing transport network, creating the necessary conditions for inward investment and business relocation through improved connectivity to local and national destinations.

Crucially, it is supporting the movement of goods and people across the Liverpool City Region, enabling them to access work, training, shopping and leisure opportunities and to be economically, socially and physically active.

A 20-YEAR JOURNEY OF CHALLENGE AND TRIUMPH

/// *Nobody gave us a chance, but we just didn't take no for an answer. We knew how important this bridge was to Halton and we were determined to deliver it.*

2004

Halton Borough Council made its final submissions to Government. Progress was stalled however, due to a combination of world events and a general election.

1994

In 1994 Cllr Tony McDermott MBE formed the Mersey Crossing Group, harnessing support from both public and private sector organisations across the region.

Local MPs, Derek Twigg and Mike Hall, Halton Borough Council and the Mersey Crossing Group worked tirelessly behind the scenes to make the case for a new crossing with key individuals across Government and sought the backing of the general public.

2000

The Government provided £600,000 towards a viability study to investigate the best options for crossing the river. This led to a consultation on possible routes with residents, local businesses and the wider travelling public.

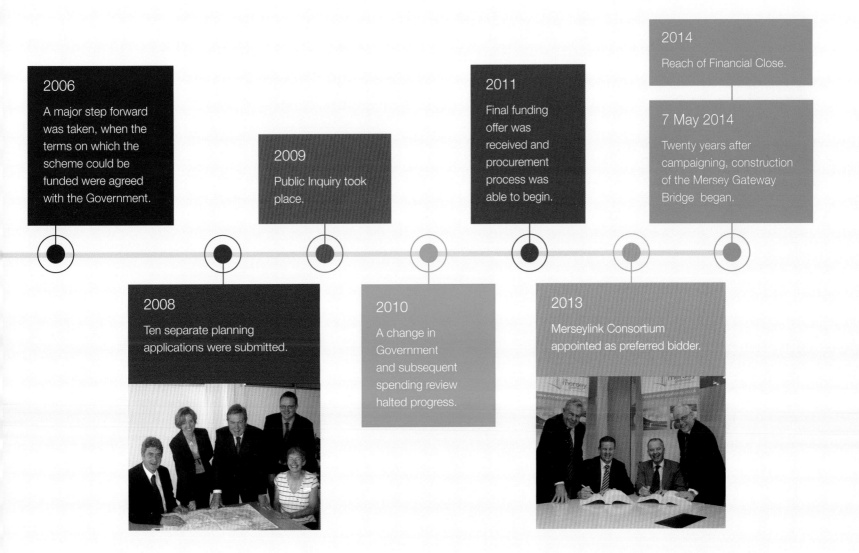

2006

A major step forward was taken, when the terms on which the scheme could be funded were agreed with the Government.

2008

Ten separate planning applications were submitted.

2009

Public Inquiry took place.

2010

A change in Government and subsequent spending review halted progress.

2011

Final funding offer was received and procurement process was able to begin.

2013

Merseylink Consortium appointed as preferred bidder.

2014

Reach of Financial Close.

7 May 2014

Twenty years after campaigning, construction of the Mersey Gateway Bridge began.

A CHANGING DESIGN AND CHOOSING THE ROUTE

The design and choice of route evolved over many years. The amount of planning involved in a project like this is phenomenal – for many people this truly has been the project of a lifetime.

From the outset there was a desire to create an iconic structure that would be a landmark for Halton and the North West. The new bridge also needed to be sympathetic to existing structures and in particular it had to complement, rather than rival, its near neighbour – the Silver Jubilee Bridge.

Initial designs for a 'second crossing' were for a two-lane, two-tier structure that included a dual carriageway that could accommodate the light rail transport system being proposed in the region around the time.

Following their appointment as lead consultants in 2001, Gifford and Partners (now part of Ramboll) began ecology, hydrology, contamination and geology studies. They also evaluated the cost, likely social and economic impacts of a new crossing and the impact on the landscape and heritage of alternative routes and different bridge types.

A tunnel was also considered but, as the most expensive option, and given that it only attracted around 40% of traffic from the Silver Jubilee Bridge, it did not meet the objectives of the scheme and was rejected.

At the start of 2003, with a bridge selected as the preferred option, the public were invited to share their views on a number of potential routes for a new crossing. In April 2003, "Route 3a" – as it was then known – became the preferred project route, meeting the objectives set by the Mersey Crossings Group – which were that together the new bridge and the Silver Jubilee Bridge would relieve conjestion, maximise development opportunities, improve public transport links and encourage cycling and walking.

The Mersey Estuary presented unique challenges for the design – stemming from the tidal character of the upper estuary; the proximity to the Specially Protected Area to the west of the Runcorn Gap; the legacy from the area's industrial history and the potential for residual contamination and the need to maintain navigation rights.

To meet these challenges, the engineers at Gifford collaborated with the Halton Borough Council team and Knight Architects to develop the Mersey Gateway 'reference design' – which at first glance is similar in many ways to the finished bridge we see stretching across the estuary today.

The reference design proposed a unique composition for a major cable-stayed bridge of three towers and two equal main spans of 300m. The shorter central pylon was a feature of this design, though the exact heights of the three pylons were changed before construction began.

All three were carefully positioned, creating the best possible engineering solution in terms of deck spans, while minimising impact within the environmentally sensitive estuary and an overall maximum height set by the approach to the nearby Liverpool John Lennon Airport.

The reference design was based on steel fabrication; however, Merseylink proposed a bridge based on a reinforced concrete deck. This is one of the areas that produced significant financial savings for the project.

Innovation also helped to deliver savings. The unique design of the 2.25km main crossing eliminated the need for intermediate expansion joints, despite the irregular plan geometry of the bridge. This has the benefits of reducing maintenance on the bridge and any traffic disruption arising from lane closures.

Elsewhere, at the Manchester Ship Canal in Runcorn, designers used scaled laboratory tests to develop a solution to protect the bridge piers in the event of an errant vessel striking the pier foundations. The design features energy absorbing "buffers" that efficiently absorb energy as they are displaced. This unique solution is believed to be a world-first and offers a cost-effective low maintenance solution to safeguard the bridge.

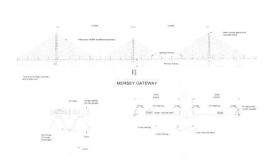

MERSEY GATEWAY

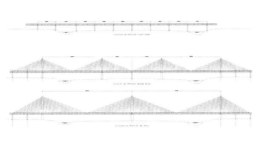

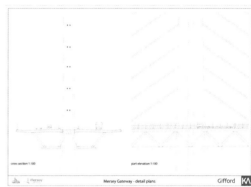

Mersey Gateway - detail plans

Gifford

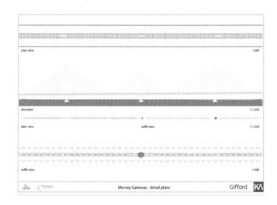

Mersey Gateway - detail plans

Gifford

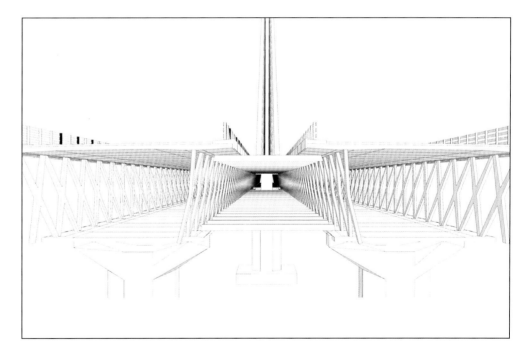

Reference design drawings and diagrams.

MINIMISING THE IMPACT ON THE ENVIRONMENT

/// The environment around Halton is unique in two very different ways. Firstly, there is a legacy of heavy industry and ground contamination and, secondly, there is the breathtakingly beautiful and very important Upper Mersey Estuary, which is incredibly important to preserve.

The Mersey Gateway is a green project which is bringing major environmental benefits to the local area. Environmental issues have been a key focus since the project was first developed and the project team has strived to ensure that the natural environment surrounding the bridge and the associated works was preserved and improved where possible.

Prior to construction, the project went through a full carbon and life cycle assessment to look at ways to minimise environmental impact.

One challenge presented early on was to contend with the legacy of Halton's chemical industry and the entire construction process was designed to ensure the team didn't disturb any deep rooted contaminants during construction.

The areas of Wigg Island in Runcorn and Spike Island in Widnes show the scale of the challenge, and frame the western part of the Upper Mersey Estuary.

While Wigg Island has been flourishing as a nature reserve since 2004, the area has a rich industrial history. During the Second World War, it was a centre for the production of mustard gas and other artillery.

Spike Island too has a rich industrial history, particularly in relation to the chemical industry. The area was once home to numerous chemical factories and rail lines and generated lots of industrial pollution.

The history of Spike Island in particular was very important to the project's environmental teams due to the legacy of contamination which remained.

A 13-month project to clean up contamination on a 5.6-hectare site at Catalyst Trade Park near Spike Island was completed so the main construction works could start.

Almost 17 tonnes of chlorinated solvent were removed from the ground on site, which far exceeded original estimates. The effort was recognised through multiple awards success for the Council and the delivery team of Ramboll and Celtic Technologies.

Elsewhere the project has processed and, where required, remediated 1,423,250 tonnes of contaminated soil made acceptable for reuse, removed 2,341,450 litres of liquid materials from the site and disposed of safely and cleaned-up 36,785,000 litres of liquid through our water treatment plant.

THE PROJECT IS FORECAST TO RESULT
IN CARBON SAVINGS OF AROUND

300,000 TONNES

Ahead of and throughout the project's construction period, the multi-award winning environmental team from Merseylink and the Mersey Gateway Environmental Trust monitored the saltmarsh on the banks of the river, its wildlife and the movement of river channels to check if they had been affected by the works. This gave an idea as to whether there had to be changes of construction methods to ensure the least possible impact on the Upper Mersey Estuary and beyond.

Ongoing monitoring shows that there have been no significant changes in the ecology of the estuary.

The team also worked to protect the local wildlife during the elements of the construction which may have affected the habitat. For example, over 600 fish including carp, perch, roach, mullet and eels were expertly rescued from one part of the St Helens Canal in Widnes to another to ensure their safety during the construction of the new bridge's north approach viaduct.

There has been extensive reuse of existing materials including carriageway pavements, which together with ground improvement techniques, reduced the need for cut and fill – the process used to move earth from one place to another to make the ground more level.

In addition, the primary construction materials – steel and reinforced concrete – are suitable to be recycled at the end of the project.

Looking forward, the environmental teams will continue their monitoring efforts and work with Tarmac, the company tasked with operating and maintaining the new bridge and road network, to continue to ensure the environment and the local natural habitats are not significantly affected by any works relating to the project.

SECURING THE FINANCE

One of the most impressive, high profile and complex projects to have reached financial completion in the UK in recent years.

The deal to finance the £1.86bn Mersey Gateway Project has been hailed one of the most innovative of recent years, setting a benchmark for the project finance sector.

An inventive combination of commercial and bank finance, the Mersey Gateway finance deal was devised by Halton Borough Council, with its advisors, KPMG and DLA Piper.

The hybrid funding solution protected public sector assets, while ensuring the deal was considered bankable for private sector investors.

From the outset, delivering an affordable scheme was the project mantra, and the deal – which saw Halton Borough Council appoint the Merseylink Consortium to design, build, finance and operate the project – delivered impressive savings of £250 million under the budget set by the UK Government in 2011.

The budget savings were in part delivered by taking a flexible approach to procurement, which allowed the Merseylink Consortium to deliver design and build innovations that secured enhanced value for money.

This unique financing deal scooped a trio of prestigious European infrastructure awards, being named Europe's Infrastructure Deal of the Year at the Project Finance International Awards;

IJ Global's European Road Deal of the Year 2014; and Partnerships Awards 2015 for 'Best Road Project'.

BIG AMBITIONS FOR WHAT COMES NEXT

The Mersey Gateway Project provides a once in a generation opportunity to transform Halton and make a major contribution to the continued growth of the wider Liverpool City Region.

The success of the land assembly programme ahead of construction paved the way for a wider regeneration strategy.

The land required for the construction of the new bridge and associated infrastructure was assembled by both agreement and the use of Halton Borough Council's compulsory purchase powers.

Over 400 parcels and plots of land were required to facilitate the construction and operation of the new bridge, together with infrastructure and highway improvements. No residential land or properties were taken as part of the land assembly process.

A dedicated land team co-ordinated accommodation works through the construction period and continue to co-ordinate the handover of residual land back to the Council for future redevelopment.

The Council revealed its ambitious vision for the area soon after the bridge opened in its Mersey Gateway Regeneration Plus Plan, which sets out a vision to deliver:

- 200-hectares of new and repositioned employment land
- 20,000 jobs (created or safeguarded)
- 3,000 new homes

The plan sets out development opportunities in eight distinct 'impact areas' across Runcorn and Widnes. These are places where the new crossing and the reconfigured network unlock land for new development.

One of these impact areas is Runcorn Station Quarter, where improvements to the road network and removal of certain structures will enhance connectivity between Runcorn Station and Runcorn town centre improving its visibility and opening up exciting new opportunities for developing the area in front of the station, creating a brilliant first impression of the town, with new places for people to live, work or enjoy.

Alongside the work of the Mersey Gateway Environmental Trust, which is creating a new nature reserve and delivering ambitious plans to conserve, protect and improve the environment, the plans show the scale of Halton's potential and its ambitions moving forward.

NEW LIFE FOR AN OLD FRIEND

AROUND 80% OF TRAFFIC IS EXPECTED TO USE THE NEW MERSEY GATEWAY BRIDGE –

MEANING THE SILVER JUBILEE BRIDGE WILL TAKE ON A NEW LEASE OF LIFE AND OFFER WALKERS AND CYCLISTS SPECTACULAR AND ENJOYABLE VIEWS ACROSS THE ESTUARY

The 'Runcorn-Widnes Bridge' was officially opened by Princess Alexandra on 21 July 1961. It was renamed the Silver Jubilee Bridge in 1977 following a widening programme, which offered a stopgap measure to cope with increasing demands for the new crossing.

With its distinctive design, the Silver Jubilee Bridge is a Grade II listed structure recognised across the North West. When opened, it was the world's third longest steel-arch span, surpassed only in size by the Sydney Harbour Bridge and Bayonne Bridge in New York.

The Silver Jubilee Bridge was closed immediately after the opening of the Mersey Gateway Bridge to enable essential maintenance of the ageing structure and to allow for reconfiguration of the deck.

When it reopens, the revitalised bridge will see the number of lanes reduced from four to two lanes for vehicles, with a new dedicated lane for cyclists.

The Silver Jubilee Bridge will be tolled, but it will be free for public transport, blue badge holders, cyclists and pedestrians.

Around 80% of traffic is expected to use the new Mersey Gateway Bridge – meaning the Silver Jubilee Bridge will take on a new lease of life and offer walkers and cyclists spectacular and enjoyable views across the estuary.

CLOSING STATEMENT
DAVID PARR, CHIEF EXECUTIVE OF HALTON BOROUGH COUNCIL

A SMALL AUTHORITY WITH A BIG DREAM

The magnitude of Halton Borough Council's achievement in delivering the Mersey Gateway Project should not be underestimated.

The Council is one of the smallest unitary authorities and the Mersey Gateway is the largest infrastructure project in the UK to be led by a single local authority.

The key to success has been an appreciation that big dreams do not come true overnight, or without the support of others. Even during the most challenging moments of the project, the belief that the bridge would one day be open to traffic never wavered.

This confidence comes from the Council's proven track record of leadership, collaboration and partnership working, particularly with private sector partners.

It acknowledges the strengths of private sector partners and the dynamism and focus on viability and delivery that is brought to the table.

An overarching philosophy is that regeneration is not just about constructing bridges or buildings and transforming the landscape. It is about changing the lives of local people by securing local employment, training and apprenticeship opportunities and boosting the local economy – as demonstrated by the Mersey Gateway.

David Parr
Chief Executive Halton Borough Council

WE HELD REGULAR WORKSHOPS AND TEAM-BUILDING EVENTS TO **IMPROVE WORKING RELATIONSHIPS,** AID COMMUNICATION AND TO IDENTIFY OPPORTUNITIES FOR INNOVATION

THE KEY TO SUCCESS HAS BEEN AN APPRECIATION THAT **BIG DREAMS DO NOT COME TRUE OVERNIGHT,** OR WITHOUT THE SUPPORT OF OTHERS

THE BELIEF THAT THE BRIDGE WOULD ONE DAY BE OPEN TO TRAFFIC **NEVER WAVERED**

PART 4:
WHO'S WHO? – GETTING THE PROJECT OFF THE GROUND

THE MERSEY GATEWAY AND PARLIAMENT
DEREK TWIGG, MEMBER OF PARLIAMENT FOR HALTON

Before becoming Halton's Member of Parliament, I had believed that we needed a second Mersey crossing. I raised the need for the Mersey Gateway during my maiden speech in the House of Commons and then went on to lead the Parliamentary case for Halton achieving the new bridge.

Achieving the Mersey Gateway to improve the lives of Halton people was always a team effort, although sometimes it seemed that the dream of achieving the Mersey Gateway was far away, we never gave up. I worked closely with former leader of Halton Borough Council, Cllr Tony McDermott MBE and the current leader, Cllr Rob Polhill gathering support for the new bridge that we knew local people wanted. I am grateful for the support received in Parliament from many Merseyside and Cheshire MPs, including Graham Evans and in particular from Mike Hall former MP for Weaver Vale.

Although we were told by Ministers that the bridge must be tolled or there would be no bridge, both I and Halton Borough Council were determined to try and get a deal from Government whereby Halton residents could travel across both our bridges toll free. I took the lead in lobbying then Chancellor George Osborne to provide additional funding on top of the Council's contribution to provide toll free travel for Halton residents.

In July 2014 after a number of months of consideration by Ministers and work with Halton Borough Council, a deal was sealed after a discussion in the House of Commons tearoom between myself and then Chancellor George Osborne.

Derek Twigg
Member of Parliament for Halton

/// *The Parliamentary campaign for the Mersey Gateway Project was a key part of the strategy to persuade Government of the need for a new road crossing and to secure Ministerial support and approval for the project. Widnes born Derek Twigg has served as the Member of Parliament for Halton since 1997.*

HALTON BOROUGH COUNCIL

/// *As sponsors of the project and ultimate owners of the Mersey Gateway Bridge, Halton Borough Council has been the driving force behind a new River Mersey crossing over the past 20 years.*

After spearheading and funding early studies and campaigning, the Council, working alongside local MPs and the Mersey Crossing Group, successfully lobbied Government, securing its backing and ultimately the funding agreement needed to take the project to market.

Halton Borough Council has always seen the Mersey Gateway Bridge as more than a piece of infrastructure, viewing it as a bridge to prosperity for the borough and beyond. With the bridge open, the Council is now focused on capitalising on the development and regeneration opportunities made possible by the Mersey Gateway. This 20-year vision is set out in its Mersey Gateway Regeneration Plus Plan.

THE MERSEY GATEWAY CROSSINGS BOARD

The Mersey Gateway Crossings Board Ltd (MGCB) is a special purpose vehicle established by Halton Borough Council with the delegated authority to deliver the Mersey Gateway Bridge Project and to administer and oversee the construction and maintenance of the new tolled crossings including the tolling of the existing Silver Jubilee Bridge.

Cllr Rob Polhill, Current Chairman and Council Non-Executive Director.

The MGCB's terms of reference and delegated authority are expressed in a Governance Agreement with the Council, set to last for sixty years.

The MGCB delivers the project on behalf of the Council, administering all contracts associated with the project and operates as a commercial (though not for profit) organisation on an arm's length basis.

Initially led by Steve Nicholson as Project Director, the MGCB currently has 14 core staff along with key technical consultants and its Board of Directors are made up of a Managing Director, Finance Director, two Non-Executive Directors and two Council Non-Executive Directors.

Core MGCB staff also manage the Mersey Gateway Environmental Trust.

MERSEY CROSSING GROUP

/ / / *In 1994, Cllr Tony McDermott MBE formed the Mersey Crossing Group, bringing together representatives from both the public and private sector.*

As the Leader of Halton Borough Council in 1994, Cllr Tony McDermott MBE recognised the importance to the project of having strong partnerships in place and of listening to a wide range of views. He formed the Mersey Crossing Group, bringing together representatives from both the public and private sector. The Group played a pivotal role in articulating a compelling business and economic case for the crossing, using their substantial influence to raise awareness with key decision makers within Government.

LIST OF MERSEY CROSSING GROUP MEMBER ORGANISATIONS

Cheshire CC, English Partnerships, Government Office North West, Halton BC, Halton Chamber of Commerce, Highways Agency, Knowsley MBC, Liverpool Chamber of Commerce, Liverpool City Council, Liverpool/Sefton Health Partnership, Merseytravel, North West Regional Development Agency, North West Regional Assembly, Peel Holdings, Sefton MBC, St Helens MBC, The Mersey Partnership, Warrington BC, Wirral MBC.

THE MERSEYLINK CONSORTIUM

The Merseylink Consortium brought together a team of world leading, international infrastructure developers to meet the needs of such a complex and iconic project, closely integrated with key international specialist contractors and suppliers.

Most of the Consortium already had substantial experience of working together, which helped to form the team to meet the exacting needs of the project. To bring new members into the team and refine the Consortium's existing working relationships, the Consortium held regular workshops and team-building events to improve working relationships, aid communication and to identify opportunities for innovation. The Mersey Gateway Project was identified as a very important project for all of the Consortium members and would be a very prestigious addition to their portfolio of top-class projects, which are spread across the world. Merseylink were delighted to be appointed Preferred Bidder and to secure Financial Close on 28 March 2014.

The Merseylink Consortium provided a financing solution which was bespoke and state-of-the-art at that time. The Consortium comprises;

BBGI SICAV

BBGI SICAV is a global infrastructure investment company listed on the London Stock Exchange with a market capitalisation of over £800 million. BBGI owns and manages 45 high-quality PPP infrastructure assets with an investment volume in excess of £10 billion and has one of the largest availability-based road transportation portfolios in the world, including three cable stay bridge projects. BBGI is committed to good governance, investing responsibly and being a good corporate citizen/long term custodian.

BBGI is proud to have been involved in the Mersey Gateway Project since construction and remains an active and committed long-term investor in this critical transportation link.

Macquarie Capital partnered with Nuveen

Macquarie Capital, the advisory, capital markets and principal investment arm of diversified financial group Macquarie, has been a pioneer and market leader in the infrastructure sector for over two decades. It has been recognised as a global leader with several industry awards, including Most Innovative Investment Bank for Infrastructure and Project Finance for two years running in 2017 and 2018, and Best Global Investment Bank in the Infrastructure Sector (source: The Banker 2017, 2018, and Global Finance 2018). With $1 trillion of assets under management, Nuveen is a premier global investment manager that has been helping clients meet their goals for more than 100 years. Nuveen's parent company, TIAA, pioneered retirement plans for teachers and non-profits, while Nuveen helped build America's infrastructure through municipal finance.

FCC Construcción (FCC) partnered with 3i Infrastructure plc

FCC Construcción is the infrastructure branch of FCC Group, which in terms of turnover and profitability is one of Europe's largest infrastructure, environmental and water services groups. In March 2014, 3i Infrastructure plc invested in the Mersey Gateway Bridge Project alongside FCC. 3i Infrastructure is a leading infrastructure investment company with extensive experience in partnering with construction and facilities management companies to design, build, finance and operate PPP projects across Europe.

CONTRACT STRUCTURE

Merseylink invited an expert group of international construction organisations which became the Merseylink Civil Contractors Joint Venture (MCCJV) – a joint venture between Kier Infrastructure and Overseas Limited, Samsung C&T ECUK Limited and FCC Construcción S.A. to design and construct the very high-quality Mersey Gateway Project assets.

MCCJV in turn secured the services of leading designers, COWI (formerly Flint Neil), AECOM, Fhecor and Eptisa, to bring the experience needed to meet the demands of such a complex infrastructure project.

Key Advisers to the Consortium were Ashurst LLP (Legal), Macquarie Capital (Financial), with CMS (Legal) supporting the joint venture.

Following Permit to Use being granted in October 2017 and the Project Road fully open to traffic, Tarmac (a CRH company), one of the UK's leading sustainable building materials and construction solutions businesses, were selected as the sub-contractor of choice to deliver Merseylink's Operations and Maintenance obligations.

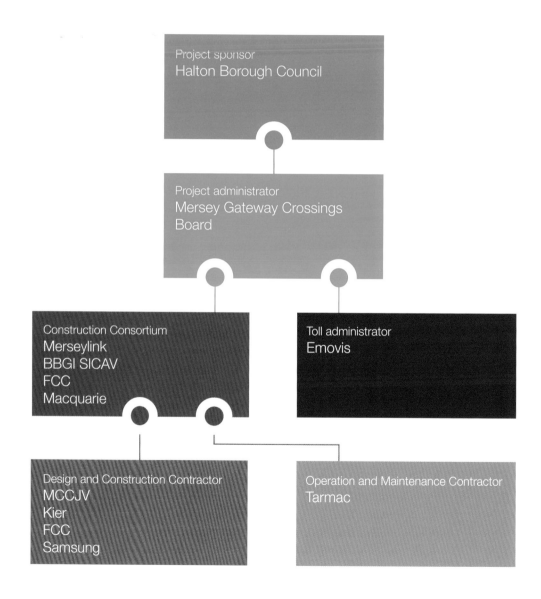

Project sponsor
Halton Borough Council

Project administrator
Mersey Gateway Crossings Board

Construction Consortium
**Merseylink
BBGI SICAV
FCC
Macquarie**

Toll administrator
Emovis

Design and Construction Contractor
**MCCJV
Kier
FCC
Samsung**

Operation and Maintenance Contractor
Tarmac

JACOBS

JACOBS

WE CONGRATULATE HALTON BOROUGH COUNCIL ON THEIR EXTRAORDINARY ACHIEVEMENT OF DELIVERING THIS AMBITIOUS AND ICONIC PROJECT.

We first joined the project as lead procurement consultant to the Council in 2006, having successfully supported the procurement of all roads projects in the UK Government's DBFO roads programme. Our experts facilitated each stage of the competitive dialogue tender process through to successful close of the transaction in 2014. At concession award, our role was novated into that of the Council's lead technical and contractual adviser to provide support during the construction and start of operation.

At Jacobs, our focus on building long-term client relationships has helped us become one of the largest and most diverse providers of technical, professional and construction services. Jacobs leads the global professional services sector, delivering solutions for a more connected, sustainable world. With a combined fiscal revenue of some $15 billion in 2017 and a talent force of more than 77,000 in 400+ locations around the world, Jacobs provides a full spectrum of services for business, industrial, commercial, government and infrastructure sectors.

DLA PIPER UK LLP

WE ARE VERY PROUD TO BE THE LEGAL ADVISERS TO HALTON BOROUGH COUNCIL ON THE MERSEY GATEWAY PROJECT.

We have worked with Halton Borough Council for many years on this project from planning and development powers, public inquiry, land assembly, structuring, public procurement and detailed negotiations through to its successful financial close in 2014. Working as a team with Halton Borough Council and its other advisers we helped run an efficient and streamlined procurement to a challenging timetable and in a manner which has delivered the project on time and with significant savings to the public sector.

It is an honour to have been involved in a project which will support the prosperity of Halton and the North West region for many years to come.

DLA Piper is a global law firm located in over 40 countries throughout the world. The North West is a great example of how we contribute to our local and global communities. Through this project as well as Albert Dock, Echo Arena and Liverpool FC's stadium redevelopment, we have played an important part in the region's dramatic physical and economic transformation. The resurgence of the mercantile spirit here, combined with a collaborative approach and willingness to push boundaries, chimes with our own approach to business: being bold, building strong relationships and delivering to the highest standards for our clients.

Liam Cowell
North West Managing Partner
DLA PIPER UK LLP

RAMBØLL

RAMBOLL

A LONG-TERM PARTNERSHIP DELIVERING LONG-TERM BENEFITS.

After 16 years working on the project it was a moment of real pride for everyone at Ramboll when the first vehicles crossed the Mersey Gateway in October 2017. In the early days as lead technical consultant we helped Halton Borough Council obtain funding for the project, sharing our engineering and environmental expertise to ensure the business case offered the best technical solution and value for money.

With a global reputation for designing and delivering major crossings and road infrastructure, Ramboll was perfectly placed to advise on options, route design, construction methods, engineering design and environmental impact assessments, and afterwards guiding the Council through technical negotiations with the bidding consortia. We were delighted to stay on during the construction phase to support the Mersey Gateway Crossings Board with the complex technical and contractual administration of the project.

The creation of the new crossing and approaches provided an opportunity to clean-up the land on either side of the Mersey, ensuring the Mersey Gateway generates a sustainable legacy. Revealing the scale of contamination on the decommissioned industrial sites, our team of environmental specialists led the award-winning advance works remediation, before advising on remediation work to be delivered during construction. On completion, our Chester office took time out to assist the Mersey Gateway Environmental Trust in removing waste plastic from the adjoining saltmarshes.

The Mersey Gateway will improve lives across the entire region. We're immensely proud of the role Ramboll has played in its delivery, and we'd like to extend a special thanks to everyone involved on the project. The shared passion and commitment, and the collaborative approach adopted by all the teams involved, turned Halton Borough Council's vision into reality.

MACQUARIE

MACQUARIE CAPITAL, THE ADVISORY, CAPITAL MARKETS AND PRINCIPAL INVESTMENT ARM OF MACQUARIE GROUP LIMITED, IS DELIGHTED TO HAVE BEEN INVOLVED IN THE SUCCESSFUL FINANCIAL CLOSE AND OPENING OF THE MERSEY GATEWAY PROJECT.

Macquarie Capital acted in many capacities on the project including as equity sponsor, adviser and debt arranger. As debt arranger, we structured the first greenfield bond wrapped by Infrastructure UK, a unit within HM's Treasury, and leveraged offshore offices to introduce a multilateral lender to the project. We committed £115 million of capital to the project across the capital structure. The project was awarded the prestigious Project Finance International's European Infrastructure Deal of the Year and Infrastructure Journal's European Roads Deal of the Year 2014.

Macquarie Capital has been a pioneer and market leader in the infrastructure sector for over two decades. Macquarie Group Limited is one of the largest infrastructure managers globally[1] and has been recognised as a global leader with several industry awards including Most Innovative Investment Bank for Infrastructure and Project Finance for two years running in 2017 and 2018, and Best Global Investment Bank in the Infrastructure Sector[2].

The Infrastructure Projects and Principal Investment group sits within Macquarie Capital and provides first class financial advisory services with flexible principal investing across the capital structure. With extensive global expertise, the group work across a broad range of infrastructure projects; from fibre optics in the US to offshore wind platforms across the UK. The group specialises in three key areas, PPPs (Public Private Partnerships), Principal Project Investment and Project Finance.

Globally Macquarie Capital focuses on six core sectors: infrastructure, utilities and renewables; real estate; telecommunications, media, entertainment and technology; resources; industrials; and financial institutions.

[1] *Willis Tower Watson Global Alternatives Survey 2017 (published July 2017)*
[2] *The Banker 2017, 2018, and Global Finance 2018*

Knight Architects

KNIGHT ARCHITECTS

KNIGHT ARCHITECTS IS PROUD TO HAVE PLAYED AN INFLUENTIAL ROLE IN THE DESIGN AND PROCUREMENT OF THE MERSEY GATEWAY.

This is a vitally important infrastructure project for Halton Borough Council and provides a distinctive new landmark for the region. We are delighted to see the crossing open to the public.

PHOTOGRAPHY CREDITS

David Hunter

Having worked in civil engineering I get most enjoyment when photographing this subject. The opportunity to capture such an amazing project on my doorstep means these images will remain very special amongst those I've taken over the years. Living locally also meant those early mornings and late nights were a little easier. I hope you enjoy the choice of images.

David Hunter November 2018
davidrhunter.photoshelter.com

Page 45: John Davidson Photos / Alamy Stock Photo
Page 96: Paul Greenhalgh / Alamy Stock Photo
Page 80: Alan Novelli / Alamy Stock Photo
Page 102: A.P.S. (UK) / Alamy Stock Photo